JD RUSSELL

Bird
of the
Blue Mountains

Margaret Baker

Robin Corringham

A Bower Bird Book

Published by:
Bower Bird Books
of Three Sisters Productions P/L
P.O. Box 4104, Winmalee, NSW 2777

Copyright © 1988 M.J. Baker and R. Corringham

Typeset by Typehouse Advertising, Katoomba, NSW
Printed by McPherson's Printing Group, Victoria

ISBN 0 9590203 5 7

First published 1988
Reprinted 1993, 2002

The authors gratefully acknowledge the ornithological advice provided by the late Graham Alcorn

Photographic Acknowledgements:
We sincerely thank the *Australasian Nature Transparencies Photo Library* and the *National Photographic Index of Australian Wildlife* for granting permission for the reproduction of transparencies from their extensive collections. The name of each photograper is followed by the list of pages on which his or her work appears.

Australasian Nature Transparencies Photo Library:
G.D. Anderson 55
J.R. Brownlie 30
J. Frazier 16, 18
K. Griffiths 40
D. Hollands 6
R. and D. Keller 12, 13, 46
I.R. McCann 29, 35, 44, 47, 53
D. and T. O'Byrne 56
F. Park 20, 42, 51
P. Röseler 5
D. and M. Trounson 8, 27
K. Vagg 26
C. Webster 17, 31

National Photographic Index of Australian Wildlife:
H. and J. Beslé 15
J. Christensen 9
J. Handel 14
W.J. Labbett 28
G. Rogerson 19
M. Seyfort 7

Additional Photographs by
R. Corringham 45
M. Dark 52
K. Griffiths 4, 10, 11, 21, 25, 32, 34, 41, 43, 48, 49, 50, 54, 57, 59
J. Purnell 22, 23, 24, 33, 36, 37, 38, 39, 58

Cover Plate: Crimson Rosella (© Ken Griffiths)

INTRODUCTION

The Blue Mountains of New South Wales, a sandstone plateau to the west of Sydney, is a temperate region ranging in altitude from 80m to 1100m. Ancient dissected landforms of grandeur and ruggedness, together with climatic variations, have given rise to a diversity of environments which include river flats, thick tall forests, low eucalypt woodland and wild open heaths. The continuing expansion of settlement on the main ridges has added to the range of habitats.

The canyons and wide valleys formed by stream erosion are clothed in warm temperate rainforest and tall open eucalypt forest, with floors of deep litter and ferns. The ridge tops and plateaus support a vegetation that has adapted to extract nourishment from poor skeletal soils to provide an unmatched array of flowering plants in open forest communities. In very rocky areas or where there is a high water table, there may be a woodland community. More specialised habitats are present in heaths of the exposed plateaus and hanging swamps at the heads of watercourses in the upper Blue Mountains. Particular bird species exploit every stratum from the deep valleys and creeks to the aerial heights and a fine balance of inter-dependence has been reached.

With the moderate depths of intersecting valleys comes a wide range of options for local seasonal and daily movements necessary to support sedentary bird populations. Great flowerings of eucalypts and nectar-laden shrubs ensure a plentiful food supply for the large number of honey-eating birds and others through the year. Spring in the Blue Mountains extends from August in Glenbrook to late November at Mount Victoria and many species follow the nectar flow.

Apart from the seasonal visitors, which may either breed here or be passage migrants, there are other birds which are seldom sighted. Cryptic birds of dense and shadowy places are indicated by their song. Others may be stragglers and vagrants blown in by bad weather or seeking refuge from drought. Still others may be aviary escapees. Cleared land, which is not a natural feature of the Blue Mountains, has encouraged a new suite of species. Up to date, introduced birds, whether exotics or local immigrants, have not interfered drastically with the endemic species. Feral cats however have decimated low-feeding birds.

The Blue Mountains National Park encircles the City of the Blue Mountains and it is largely due to this vast undisturbed natural area that there is a familiarity with wild birds in the urban areas. Some famous bird watching places include Blue Gum Swamp at Winmalee and Braeside Walk at Blackheath.

With 187 species recorded, the Blue Mountains' birds are among the richest and most colourful in the world. Presented in this book are 56 of those most commonly seen and heard. A checklist of all sighted species is included for those who wish to investigate further.

Maned Duck
Chenonetta jubata

Appearance Length about 480mm. The male has a brown head and neck with a mane of black feathers, a black lower back and a mottled fawn breast. Wings are grey with black tips and a turquoise speculum. The female has grey-brown plumage with mottled brown pale breast and flanks. Both sexes have a short black bill, olive-brown legs and brown eyes. A white line extends above and below the eye of the female.

Voice An extended *'a-ark'*.

Breeding In spring. The nest is a hole in a living tree, sometimes high, sometimes distant from water. The nine to twelve cream or white eggs are covered by the parents with down.

Behaviour A wary bird, it will flatten its body to look like dead wood if alarmed, particularly when with young. It is seen as often on the ground or perched on logs and branches as on the water. Grazing during the night at a selected site, it returns to a waterside camp for the day.

Food Terrestrial vegetation particularly grass.

Distribution The Maned or Wood Duck is most common in the moister parts of south-eastern Australia. Though there is some seasonal movement, it is a resident breeder in the Blue Mountains, where it frequents the margins of swamps and parks as in the vicinity of Wentworth Falls Lake.

Maned Duck

Wedge-tailed Eagle *Aquila audax*

Appearance — Length 980mm. Sexes are similar though the female is larger. The birds are an overall black, developing brown tonings on the nape and wings with age. The long tail is wedge-shaped and the wing span exceeds 2.5m. The strong bill is grey with a yellow patch and a black hooked tip. Eyes are brown, the feathered legs black and the clawed feet grey. Immatures are layered browns above and darker on the wings. Brown under part feathers are cream edged.

Voice — A weak whistle is heard particularly during mating. Loud screechings are uttered if disturbed when on the nest.

Breeding — June to August. The nest is a large platform of sticks lined with leaves, placed in the fork of a tree. The one to three large eggs are cream with brown and purple spots.

Behaviour — This is Australia's largest bird of prey. It hunts singly, in pairs or in small parties. During the day it spends much time soaring and gliding in slow graceful displays often high above its territory.

Food — Carrion, birds, reptiles, wallabies, rabbits and sick lambs.

Distribution — Widespread throughout Australia from timbered areas to arid plains. A sedentary breeder in the Blue Mountains it may be seen soaring over the Megalong and Jamison Valleys and at Kanangra Walls.

Wedge-tailed Eagle

# Brown Falcon	*Falco berigora*

Appearance	Length about 500mm. Sexes are similar, though the female is larger than the male. Overall colour varies markedly from light to dark brown, the throat and under parts being paler with black streaks. They sometimes look pale gold when in flight. A black moustache lines the paler cheeks. The bill is metallic grey, eyes brown and legs blue-grey. Immatures are dark brown above and buff coloured below.
Voice	A loud cackling call.
Breeding	June to November. The nest may be a tree hollow, the recycled nest of another hawk or a new one built of sticks. The two to five buff eggs are darkly blotched.
Behaviour	The hunting techniques of the Brown Falcon vary from other species. It pounces on ground prey after watching from a high vantage perch or while hovering, rather than making chase. Its flight is relatively slow. Couples perform aerial acrobatic displays before breeding.
Food	Mammals, reptiles, carrion, birds and insects.
Distribution	This species is widespread in forested and open country throughout Australia. In the Blue Mountains its numbers have been favoured by clearing, particularly in the pastoral valleys of Kanimbla and Megalong. While partially migratory it is a resident breeder here.

Brown Falcon

Brush Bronzewing *Phaps elegans*

Appearance	Length 290mm. The male is chestnut-brown on the forehead, nape and throat, with a brown back and upper tail. The brown wings have bronze, blue and green banding. Under parts are grey. In the female the chestnut-brown colouration is restricted to a smaller throat patch and a line behind the eye. Eyes and bill are dark brown and the short legs and feet are pink. Immatures resemble duller females.
Voice	A low sad repetitious *'coo'*.
Breeding	From October to January, though it can be in most other months. The shallow nest of loosely packed twigs is built close to the ground in a shrub or fallen branch. The two small white eggs are smooth and glossy.
Behaviour	Single birds or pairs feed quietly on the ground, generally under the shelter of bushes from which they will fly upwards with a loud clatter if disturbed.
Food	Seeds, particularly those of acacias, and fruit.
Distribution	Mostly found in coastal parts of the south-east and south-west of the continent. In the Blue Mountains they prefer the shrubby understories of eucalypt forests, though clearance of bushland has reduced their numbers here. Being ground-feeding birds they are also easy prey to feral cats and foxes.

Female Brush Bronzewing

Yellow-tailed Black Cockatoo   *Calyptorhynchus funereus*

Appearance	Length about 650mm. Both sexes have funereal black plumage, the feathers edged with yellow, and a broad yellow tail band. Eyes are dark brown and legs greyish. The male has a dark grey bill, a small pale yellow cheek patch and a pinkish ring around the eye. The bill of the female is lighter grey, the cheek patch larger and more yellow, and the eye ring dark grey.
Voice	A drawn out creaking wheeze, a harsh screech and chuckles.
Breeding	Spring to summer. Nest is a hollow in a high tree, lined with woodchips chewed from the walls. Though two white eggs may be laid only the first of the young survives.
Behaviour	Generally moving in family or small groups, speed over distance is comparatively rapid, though wing movement is slow. Some seasonally join large migratory flocks, moving to lower altitudes after breeding. Juveniles will keep up a persistent wheezing for hours while camped in one tree. The powerful beak is used to tear apart timber and seed cases in search of food.
Food	Exotic pine seeds, woody-cased seeds of hakeas, banksias and *Lambertia formosa,* and wood-boring larvae.
Distribution	In the forests of south-eastern Australia. They are breeding residents of the Blue Mountains, where they are greatly attracted to the abundance of exotic pines.

Yellow-tailed Black Cockatoo

Gang-gang Cockatoo *Callocephalon fimbriatum*

Appearance	Length 350mm. The male is an overall grey with the feathers scalloped white and the crested head red. The female is similar to the male but with more yellow in the breast feathers and the crested head grey. The eyes are dark brown and the bill and legs light grey. Immatures resemble the females though the young males have faint red on the head and crest.
Voice	Wheezy creaking in flight and low growling when feeding.
Breeding	October to January. Nest is a hole chewed to size, in a decaying eucalypt. The two eggs are white and rounded.
Behaviour	It is found in family or larger flocks which have a habit of sequentially exhausting individual food sources. Sitting amongst the foliage they hold seed pods in their claws, noisily cracking them open with powerful beaks. They indulge in mutual preening, often during the hottest days. Despite large wing areas they are inactive birds.
Food	Seeds of native and introduced plants, including wattles, eucalypts, conesticks and hawthorn.
Distribution	From the coast to the ranges of south-eastern Australia. Though resident breeders in the eucalypt forests of the Blue Mountains, during autumn and winter they move to woodlands and gardens, following food supplies.

Male Gang-gang Cockatoo

Galah
Cacatua roseicapilla

Appearance — Length about 360mm. Sexes are similar. The under parts and neck are deep pink, the crown and back of the head pale pink and the back, wings and tail light grey. The eyes of the male are brown, the female's eyes are pink and the eye-ring colour varies. The bill and legs are grey.

Voice — Has a variety of calls from single notes to harsh screeching territorial cries. Flock noise is tremendous.

Breeding — August to November. Nest is a deep hole in a tree or a hollow limb lined with green leafy twigs. The two to six eggs are small, white and oval.

Behaviour — Galahs are commonly seen in small flocks scavenging along the Blue Mountains highways. During the breeding season they separate into pairs and return to their habitual nesting tree. The male clears a bare patch from the entrance to the nesting hollow. Perhaps to establish ownership, the breeding pair often wipe their bills and faces over this area.

Food — Ripened seed gathered from the ground.

Distribution — Widespread throughout Australia except along the extreme east coast, south-west and parts of Tasmania. The species is an immigrant to the Blue Mountains and, preferring open country, it frequents inhabited parts, particularly where there are cleared grassy areas with access to water.

Galah

Sulphur-crested Cockatoo *Cacatua galerita*

Appearance Length about 490mm. The sexes are similar. Plumage is white, washed with yellow on the undersides of the broad wings and tail feathers. The narrow yellow crest extends forward. Eyes are black with bill and legs dark grey.

Voice A vocal bird, it has a harsh squawking contact call and an alarm call which is a succession of screeches.

Breeding August to January in the Blue Mountains. Nest is a hollow deep in a tall eucalypt. The two eggs are white.

Behaviour This is a bird of flocks which are noisily active in the cooler parts of the day. When ground feeding all birds may suddenly rise into the air having been warned of an intruder by the screechings of a few birds perched in the trees above. It has a slow measured flight.

Food It is generally a ground foraging species of grains, bulbs, insects, wood-boring larvae and, in the Blue Mountains, Radiata Pine seeds. It exhausts food supplies of each area.

Distribution A common bird of most types of forested country it has a widespread distribution from south-eastern Australia northwards across to the Kimberleys of Western Australia. An abundance of exotic conifers and permanent water in the Blue Mountains has benefited the establishment of the species here, probably from aviary escapees.

Sulphur-crested Cockatoo

Australian King Parrot *Alisterus scapularis*

Appearance	Length 430mm. The male has a bright red head and under parts. The back is green. Wings are green with a light green band. The rump is blue and the tail is black. The upper mandible is red with a black tip. The female has a green head, back and breast and red belly. The tail is green to blue-green. The bill is a grey-brown. Male and female have yellow eyes and grey legs. Immatures are similar to the adult female until, at about 30 months, males acquire their adult plumage.
Voice	A shrill squeak contact call, and a high pitched piping.
Breeding	Spring to January. The nest, which is lined with wood dust, is deep in a eucalypt hollow. The three to five eggs are white.
Behaviour	King Parrots are wary birds which move in small families. Their flight through trees is fast and manoeuvring but measured and heavy over distance flying.
Food	Fruit, seed, nectar and blossoms of native and introduced plants.
Distribution	The eucalypt forests of eastern Australia. In the Blue Mountains there is some local migration in spring to the higher plateaus, otherwise it is a sedentary bird and resident breeder. It is common in gardens.

Male Australian King Parrot

Crimson Rosella *Platycercus elegans*

Appearance	Length 360mm. Sexes are similar though the male is a little larger. The body is bright crimson with bright blue cheek patches. The wings are blue and black and the tail is blue. Eyes are brown, the bill pale grey and the legs and claws darker grey. Immatures are an olive-green with a patch of crimson on the forehead and throat. They assume adult plumage at about sixteen months.
Voice	A vocal bird with a variety of calls including bell-like songs of from three to five notes, repeated short squawks and excited chatter in company.
Breeding	September to January. The nest is in a tree hollow. The three to five eggs are white.
Behaviour	An animated bird whose swift flight through trees becomes undulating in the open. Its tail fans on alighting. Immatures band together in flocks.
Food	Seeds, particularly those of eucalypts and wattles, insects, cultivated fruit and sometimes eucalypt blossoms.
Distribution	Widespread in the eucalypt forests and rainforests of south-eastern Australia wherever there are mature trees with nest holes. It is a very common species in the Blue Mountains where it is a resident breeder. It is easily enticed into gardens.

Crimson Rosella

Fan-tailed Cuckoo
Cuculus pyrrhophanus

Appearance	Length 275mm. The sexes are similar. Upper parts are a shiny dark grey and the throat, breast and belly are a dull rufous. On the tail feathers dark-blue 'diamonds' alternate with whitish notches. Under parts of the female are duller. A yellow eye-ring surrounds the brown eye, the bill is black and the legs yellowy-brown. Immatures are very different. Upper parts are brown with the feathers edged red-brown. Under parts are mottled brown, grey and black.
Voice	A repetitious sad descending trill and a single whistle.
Breeding	August to December. A parasitic bird, it may place its eggs in the dome-shaped nests of small thornbills or in the open nests of honeyeaters. The one small whitish egg has a band of brownish spots at the larger end.
Behaviour	A bird with easy undulating flight, it will dart swiftly to catch an insect that it has detected while sitting quietly on a low branch, then fly to another perch to eat it. It is also seen hopping along the ground on its short legs.
Food	Caterpillars and insects mainly taken from low foliage.
Distribution	Common throughout the timbered parts of eastern and south-western Australia from the coast to the semi-arid zone. In the Blue Mountains it ranges from rainforests to suburban gardens.

Fan-tailed Cuckoo

Channel-billed Cuckoo *Scythrops novaehollandiae*

Appearance	Length 600mm. Sexes are similar. The body is pale grey with a white belly. Back and wings are darker grey and the long tail is barred dark brown. The eyes are red, the long bill white and the legs grey. The upper parts of immmatures are more buff coloured, the eyes brown and the bill reddish.
Voice	It has a loud harsh cry, often uttered in flight and, like some other cuckoos, calls through the night. Its call, heard from September to January, could be mistaken for that of the Koel Cuckoo.
Breeding	October to December. A brood parasite, hosts include currawongs and crows, whose nests may receive eggs from more than one female Channel-bill. The two or more eggs are white to cream with lavender and brown markings.
Behaviour	The Channel-billed Cuckoo is usually solitary but may be in pairs. It generally flies high over the trees. If observed by potential host species it is severely harassed and routed. Young currawongs will join adults in the chase.
Food	Insects and native fruits.
Distribution	In suitable areas of food and hosts in the eucalypt forests of northern and eastern Australia. It is a spring migrant from New Guinea and Indonesia. In the Blue Mountains it breeds mainly in the middle and lower altitudes.

Channel-billed Cuckoo

Boobook Owl
Ninox novaeseelandiae

Appearance	Length 340mm. This is the smallest Australian owl. Sexes are similar. The upper parts are brown and the wings and back may be spotted white. Under parts are streaked and mottled with white and light brown. There is a dominant dark mask behind the forward looking yellow eyes. The bill is blue-grey. The grey feet are well adapted to grasping prey.
Voice	The distinctive call, heard particularly on warm nights, is a measured two note descending *'boo-book'*. This may be heard as *'mopoke'*, providing the bird with its alternative common name of Mopoke.
Breeding	September to November. Nest is a tree hollow lined with decayed wood dust, located from 1-20m above the ground. The two to four eggs are dull white.
Behaviour	Nocturnally active, its light wing-loading and soft feathers permit silent night flight. This bird roosts by day in thick foliage and makes itself inconspicuous when disturbed. If discovered by small birds it is harassed unmercifully.
Food	Mainly insects, small mammals and birds.
Distribution	This, the most common Australian owl, is found wherever nesting sites are available, preferably in forests. It is a breeding resident in the Blue Mountains.

Boobook Owl

Tawny Frogmouth *Podargus strigoides*

Appearance	Length 330-460mm. Though colour varies it is an overall shadowy grey, mottled and streaked with greys and brown. The male is usually greyer and smaller. Eyes are yellow and the strong hooked wide-gaping bill is surrounded by long grey bristles. The feet are comparatively weak.
Voice	A mechanical low pitched call of repeated notes, *'oooo'*
Breeding	August to December. Nest is a platformed collection of sticks built across the fork of a horizontal branch. The two to three eggs are white. Tawny Frogmouths may mate for life and use the same nest site year after year.
Behaviour	Being a nocturnal bird, the Tawny Frogmouth is often mistaken for an owl. It is more closely related to the kingfishers and swifts. Active particularly at dusk, it flies silently and slowly on short broad wings to capture moving prey. The camouflage of its diurnal roosting position, which is often against an upright trunk, makes it almost invisible.
Food	Insects and small animals moving on the ground.
Distribution	This is an almost Australia-wide species, being found wherever suitable trees for roosting and nesting grow. Resident and common in the timbered parts of the Blue Mountains, it shows a preference for eucalypt woodland.

Tawny Frogmouth

Laughing Kookaburra *Dacelo novaeguineae*

Appearance	Length 460mm. Sexes are similar. A dark line runs through the brown eye to the back of the buff to off-white head and neck. The lower back is dark brown and the white-tipped tail is barred with black and reddish-brown. The brown wings are scalloped blue. Males sometimes have a greenish-blue patch on the rump. The powerful bill is black above and grey below. The legs and feet are grey. Immatures have a short black bill and blue-tipped tail.
Voice	The famous '*koo-koo-koo-koo-ka-ka-ka-ka*' is chorused through the bush, typically at dawn and dusk. Other short calls including chuckles are also typical.
Breeding	September to January. Nest is a large bare hole in a tree trunk. The one to four eggs are white and rounded.
Behaviour	Most of the familiar calls and antics of these sedentary birds are associated with the advertising and defence of permanent territories. Family groups of pairs and young adults are particularly vocal before the breeding season. Heads and tails are raised when 'laughing'.
Food	Small snakes, lizards, worms, centipedes and insects.
Distribution	Widespread in the woodlands and open forests of eastern Australia. They are opportunistic ground feeders in areas of land clearance in the Blue Mountains.

Laughing Kookaburra

Sacred Kingfisher *Halcyon sancta*

Appearance Length 210mm. This is a stunning bird coloured turquoise green on the head and shoulders and bright turquoise blue on the lower back, wings and upper tail. Beneath the tail is light grey. A cream wedge extends from the brown eye to the black bill. A black and cream collar runs from behind the eye and throat. The lower under parts are buff to light brown. Feet are blackish. The female tends to be duller, greener and slightly larger. Immatures are browner.

Voice A monotonous *'ke-ke-ke-ke'* marks the breeding season.

Breeding September to December. Nest is a hollow in a tree limb or earth bank. The three to six oval eggs are glossy-white.

Behaviour Sacred Kingfishers are migratory birds, arriving for the breeding season from islands north of Australia and leaving again in March. Pairing only for breeding, the family group exchanges calls to establish its territory. Both sexes excavate the nest site, using bills for digging and feet to remove debris. During the day this bird watches for its prey from a low branch where it sits very still, sometimes bobbing its head.

Food Ground dwelling insects, insect larvae and small skinks.

Distribution Widespread in Australia apart from the arid interior it is seasonally common in eucalypt forests and woodlands.

Sacred Kingfisher

Superb Lyrebird *Menura novaehollandiae*

Appearance	Length 900mm. Upper parts of the male are dark brown and the under parts are lighter. The 550mm long tail, brown above and silver-grey below, has 2 long narrow feathers, 12 filamentous feathers and 2 outer banded lyre-shaped feathers. The female lacks the lyre-shaped tail. Its tail is bunched with the feathers slightly twisted. Eyes are black, and the bill, legs and powerful feet greyish.
Voice	The resonant song, mainly heard in the cooler months, is loud, incorporating its own calls with the superb mimicry of local bird sounds. The alarm call is a high shriek.
Breeding	Mainly from June to July. Nest is a side-entranced, bulky dome of sticks and ferns. In the Blue Mountains it is often sited on a rock ledge. The one grey to purple-brown egg is darker blotched.
Behaviour	A secretive ground bird, it spends most of its time scratching through the leaf litter over considerable areas. Though awkward in flight it roosts in trees. The male performs an extravagant and vocal courtship display, its shimmering extended tail thrown forward over its head.
Food	Invertebrates and seeds on the ground.
Distribution	Common in south-east Australia, it is a resident breeder in the rainforests and forested valleys of the Blue Mountains.

Male Superb Lyrebird

Welcome Swallow *Hirundo neoxena*

Appearance	Length about 150mm. Sexes are similar. The upper parts are blue-black, with the forehead and throat rufous, wings duller and belly grey. The dark brown tail is forked. The eyes, bill and legs are dark brown.
Voice	Sweet twittering and warbling.
Breeding	August to December. More than one brood is produced in a suitable season. The cup-shaped nest of grass and mud is attached to a tree hollow, cave wall or man-made structure. The four to five white eggs are marbled purple-brown.
Behaviour	Though generally seen as a lone bird or breeding pair, it is sometimes a colonial species. Its flight is swift graceful and darting. Before the breeding season each bird appears to return to the same nesting site, or one close by, with the same partner.
Food	Aerial insects taken on the wing from among the tree tops.
Distribution	Widespread in eastern and southern Australia in the lighter forested areas and towns. The Welcome Swallow breeds in the Blue Mountains where, favoured by the reduction of forest and an increase in potential breeding sites in buildings and bridges, its numbers have increased with the spread of settlement. Partially migratory, some move to warmer areas in winter.

Welcome Swallow

Black-faced Cuckoo-shrike *Coracina novaehollandiae*

Appearance	Length 330mm. Sexes are similar. The back and breast are pale grey and the forehead, face, throat, white-tipped tail and flight feathers, eyes, bill and legs are black. The belly is white. Immatures have less black on the face.
Voice	A rolling *'chur'* in flight and a slightly drawn out *'cherur-cherur'* when perched.
Breeding	August to February. The nest is a small shallow saucer of sticks and bark bound with cobweb. It is almost invisible in a tree fork a few metres above the ground and is so small that the young are sometimes blown out during strong winds. The two or three olive-toned eggs are plain or marbled with grey and brown.
Behaviour	The Black-faced Cuckoo-shrike has a distinctive flapping then gliding flight and shuffles its wings after alighting. It moves seasonally in pairs or small flocks. Breeding pairs actively defend nestlings, diving and snapping at intruders.
Food	Insects, particularly cicadas, and larvae, taken from the ground, foliage or air, small fruits, berries and seeds.
Distribution	Australia wide, it is common in open forest and woodland and is often seen in parks and gardens. As in other parts of south-east Australia, it breeds in the Blue Mountains then migrates north in autumn, though a few birds winter here.

Black-faced Cuckoo-shrike

Rose Robin   *Petroica rosea*

Appearance	Length 110mm. The male is dark grey above with a white spot on the forehead. The throat is dark grey and the roseate breast shades to a white belly. The female is grey above with a dull forehead spot. The breast is grey, sometimes tinted rose. The eyes, bill and legs are dark brown.
Voice	A faint thin *'tic-tic-tic-tic-tic'* followed by a churring *'chew-chi-chew'*.
Breeding	September to February. Up to three broods are produced in a suitable season. The cup-shaped nest is made of fibre and moss, lined with softer material, camouflaged with lichen and set on a look-alike branch fork high above the ground. The two or three bluish eggs are speckled purple-brown.
Behaviour	This is an engaging and inquisitive little bird which flies about gracefully. It frequently forages high in the trees or, with tail fanned, darts after flying insects. While perched it will drop its wings and lift its tail, or twitch both as though to take off.
Food	Spiders, insects and larvae from the ground or in foliage.
Distribution	Widespread in the wetter forests of the coast and ranges of south-eastern Australia. Resident breeders in the open forests and rainforest of the Blue Mountains, they migrate from the valleys to the ridges in winter.

Male Rose Robin

Scarlet Robin *Petroica multicolor*

Appearance	Length 130mm. For the male the upper parts, throat and head are black and the forehead conspicuously white. A ragged white stripe along the outer wing forms a v-shaped band on the back. The breast is scarlet. The upper parts and throat of the female are grey-brown, the forehead and wing stripe white and the breast red. Juveniles are similar to the female but duller. Though sometimes confused with the Flame Robin, the Scarlet Robin is distinguished from that species by the red breast of the female and the conspicuous white brow and black throat of the male.
Voice	Sweet quiet trilling notes particularly at dawn.
Breeding	July to December. The cup-shaped nest is made of bark, grasses and moss bound with cobweb and set in a forked tree branch. The three or four eggs are greenish-white with dark shadowing towards the larger end.
Behaviour	Found singly or in pairs, this is a lively bird which flicks its tail on alighting and while perched. It joins mixed small bird flocks out of the breeding season.
Food	Small insects. Its foraging levels vary seasonally.
Distribution	This species inhabits the drier forests of south-east and south-west Australia. It is a resident breeder in the Blue Mountains, moving from valleys to the ridges in winter.

Male Scarlet Robin

Eastern Yellow Robin *Eopsaltria australis*

Appearance	Length 150mm. Sexes are similar. The upper parts are grey-brown, the chin white and the under parts yellow. Eyes and legs are brown and the bill black.
Voice	A pleasant clear piping note repeated three or four times continuously. It is particularly vocal in spring.
Breeding	June to February with up to three broods. Nest is a cup made of grass and bark bound with cobweb and lined with softer material. It is often constructed some distance from the ground in the fork of a slender tree. The two or three eggs are pale green to blue with dark reddish markings.
Behaviour	The Eastern Yellow Robin moves quietly and is friendly and trusting. It is most often seen alone but does form small family groups. Though among the early risers it is still hunting after dusk. It clings sideways low on a tree trunk to watch for ground prey, darts in quickly to grasp the victim in its bill, then returns to its vantage point. It is often the first visitor to burnt areas and will even hunt close to flames.
Food	Generally ground insects, larvae and spiders. It occasionally forages in foliage.
Distribution	Common in the better-watered open forests of eastern Australia, it is a sedentary resident of the Blue Mountains.

Eastern Yellow Robin

Golden Whistler *Pachycephala pectoralis*

Appearance	Length 170mm. The head of the male is black, back of the neck yellow, and other upper parts olive-grey. The white throat is separated from yellow under parts by a black band. The female is an overall brown-grey. Immatures are similar to females. It is distinguished from the Rufous Whistler by the rufous breast and belly of the latter.
Voice	A sequence of two to four or more noted whistles ending with an ascending note. It calls continually in the breeding season. A soft indrawn whistle is heard through the year.
Breeding	September to January. The nest is a bowl of plant stems and pliant twigs, lined with finer material, in a tree fork or dense shrub. The two or three creamish eggs are speckled.
Behaviour	A quiet-moving bird it is seen singly or in pairs. It will sit on a branch bobbing its head from side to side seeking insects from the foliage.
Food	Insects and spiders from trees and occasionally the ground.
Distribution	Found from the coast to semi-arid areas of eastern and south-western Australia wherever there is unbroken canopy. It is a resident breeder in the Blue Mountains, with some seasonal, altitudinal migration. There is some overlap of its range with that of the Rufous Whistler but that species prefers drier areas and is a summer migrant here.

Male Golden Whistler

Rufous Whistler *Pachycephala rufiventris*

Appearance	Length 170mm. The male has dark olive-grey upper parts, a white throat and a black band above a rufous breast and belly. Wings and tail are brown-black and the bill is black. The plainer female has brown streaked, pale under parts. The bill is brown. Immatures are indistinctly coloured and similar to adult females.
Voice	A vocal bird, both male and female contribute many calls and songs. The identifying song is a rapid succession of single notes ending with a stressed *'ee-chong'*. A sudden loud noise may provoke song.
Breeding	From September to February with one or more broods. The nest is a twiggy cup placed in a bushy tree fork or shrub. There are two or three dull olive eggs with darker blotches. Birds mate for life and continually use the same territory.
Behaviour	They generally move quietly in pairs through the upper foliage, sometimes joining mixed flocks particularly during the autumn migration.
Food	Mainly insects, spiders and sometimes fruit.
Distribution	Throughout most of Australia but avoiding the denser forests. The eastern populations migrate to the north in autumn, returning in the spring to breed. Some birds overwinter in the Blue Mountains.

Male Rufous Whistler

Grey Shrike-thrush *Colluricincla harmonica*

Appearance	Length 230mm. Upper parts of the male are grey, though sometimes a dark grey crown and lower back are separated by a brown upper back. Under parts are light grey with a darker breast. A white streak extends from the dark brown eye with a dark eye-ring to the black bill. Throat and breast of the female are lightly streaked, the eye-ring white, bill brown and cheek line grey. Flight feathers of immatures are rufous edged and the breast is sometimes brown streaked.
Voice	Well known as a songbird, the male emits rich phrases with rising endings during the breeding season. There is also a repeated penetrating whistle and a grating *'oick'*.
Breeding	July to February. A bowl-shaped nest of bark and coarse grass is built in an upright fork of a shrub or tree, or in a crevice in a stump, rock ledge or building. The three or four small white eggs are darker speckled.
Behaviour	The Grey Shrike-thrush is usually seen alone or in pairs vigorously hopping along branches, through undergrowth or on the ground in search of food.
Food	Insects, skinks, small mammals and nestlings.
Distribution	Widespread in Australia from rainforest to woodland and around gardens and houses. In the Blue Mountains it leaves the ridges in summer, prefering the cooler valleys.

Grey Shrike-thrush

Rufous Fantail *Rhipidura rufifrons*

Appearance	Length 160mm. Sexes are similar. The upper parts are a brownish grey, darker towards the lower half of the tail. The rump and upper tail are a contrasting bright orange-rufous. This colour is repeated on the forehead and eyebrow. The white throat is edged below with black. The breast feathers are black, edged with white and the belly is cream. Eyes are dark brown and the bill and legs grey. Immatures are more rufous above and browner below.
Voice	A single high-pitched note or thin song.
Breeding	Late October to February. The nest is a short-tailed cup of cobweb-bound fine grasses and bark. It is generally built near water in the fork of a shrub or tree up to 10m above the ground. The two or three eggs are yellowish with brown and grey spots.
Behaviour	The Rufous Fantail's apparently weak flight is marked by a permanently fanned tail.
Food	Insects taken on wing and from low foliage.
Distribution	Along the northern and eastern fringes of the continent. Migratory, it arrives in the Blue Mountains by mid-September and the adults leave for northern Australia in March. Juveniles follow in April. It breeds in damper forests then moves out into any area with low shrubs.

Rufous Fantail

Grey Fantail

Rhipidura fuliginosa

Appearance	Length 160mm. Sexes are similar. The upper parts are grey-brown with a white band over the eye and another over the ear coverts. The throat is white. Under parts are buff and the long tail has white outer feathers. The eyes, bill and legs are black. Immatures are duller with buff markings instead of white.
Voice	A small voice with a repeated *'chip-chip'* and a high pitched ascending melodic song.
Breeding	August to January. The nest is a fibre wine glass bound with cobweb to a thin horizontal tree fork, the 'stem' hanging 150mm. The two to four eggs are dull cream with brown markings.
Behaviour	There is a continually active acrobatic pursuit of flying prey at all levels of the vegetation. The tail is almost permanently fanned. Even when it perches the bird moves restlessly from side to side.
Food	Flying insects mainly from the air, but also a proportion from the ground and on foliage.
Distribution	Enjoys most habitats in Australia except for the arid interior. It is a resident breeder in the Blue Mountains. Out of the breeding season it moves in mixed small bird flocks and is commonly seen in gardens in the cooler months.

Grey Fantail

Eastern Whipbird *Psophodes olivaceus*

Appearance Length 260mm. Sexes are similar. Head and crest are black, upper parts olive-green and the tail white-tipped. The throat is black with white sides. The black breast is mottled white and the belly is grey. The eye is brown, the bill black and the legs red-brown. Juveniles are olive all over.

Voice Is a rising whistle, then the male 'whip cracks' and the female utters a quick *'chew-chew'*. Sometimes the male provides these last two notes. It has many other beautiful short songs all on the 'whip' theme.

Breeding October to December. The flattish cup-shaped nest of long twigs lined with finer material is generally constructed close to the ground in dense undergrowth. The two small pale blue eggs have black and greyish markings.

Behaviour A shy sedentary bird, its presence is often only detected by the distinctive 'whip crack' call. Like the Lyre Bird it uses its strong feet to churn through leaf litter for insects. In reluctant flight it will fan its long tail to reveal the whitish tips, but trails it when moving across the ground.

Food Insects and their larvae from litter and dense undergrowth.

Distribution Widespread through coastal eastern Australia from the rainforest to drier eucalypt forests, it is a resident breeder in the Blue Mountains.

Eastern Whipbird

Superb Fairy-wren *Malurus cyaneus*

Appearance Length 160mm. The plumage colour of females and non-dominant males is similar out of the breeding season. Upper parts are brown and under parts cream. The female has a greeny-blue tail and chestnut-brown bill and legs. The male has a bluey-brown tail, black bill and legs, and brown eyes. During the breeding season the male develops a turquoise blue crown, upper back and cheek patches. The nape of the neck and lower back are black. The tail is deep blue. Dominant males retain their breeding plumage throughout the year. Young males gain blue tails in their first autumn and full breeding plumage the next spring.

Voice A noisy chitter.

Breeding September to January. Nest is a domed ball of grass and twigs woven with cobweb and built in dense foliage near the ground. The two or three white eggs are red spotted.

Behaviour Family groups which include several adult males are often seen darting through low dense undergrowth. They briskly forage in clearings and shrubs over sizeable territories.

Food A variety of small insects.

Distribution South-eastern Queensland to South Australia wherever there is dense undergrowth. They are not so common now in the Blue Mountains as they are easy prey for cats.

Male Superb Fairy-wren

Variegated Fairy-wren *Malurus lamberti*

Appearance	Length 140mm. The head of the male is blue, nape and rump are black and the belly white. The cocked tail is dull blue. Its distinctive mark is a bright chestnut shoulder patch. The female is grey-brown with a blue washed tail and dull red about the eye. Immatures are similar to adult females. The male loses its blue plumage out of the breeding season. This species has a lighter build and a longer tail than the Superb Fairy-wren.
Voice	A full trill.
Breeding	Generally communally from September to January with successive broods in suitable conditions. Nest is a side-entranced bowl made of grasses and spider egg sacs, arranged in low dense bushes or grass tussocks. The three or four white eggs are speckled red-brown.
Behaviour	The birds live in groups of from two to five. They are strongly territorial but forage further afield. There is some overlap in breeding territories with the Superb Fairy-wren.
Food	Invertebrates such as weevils and caterpillars, and seed.
Distribution	There are several races in this species which occupy different habitats through nearly all of Australia. The Blue Mountains birds are resident breeders in most environments except rainforest and urban areas.

Male Variegated Fairy-wren

Rock Warbler *Origma solitaria*

Appearance Length 140mm. Sexes are similar. Plumage is an overall brown, shading to rufous on the rump and cinnamon on the face. The throat is speckled white. Eyes are red-brown and the bill and legs dark brown. Immatures are paler versions of the adults.

Voice A shrill sharp *'pink'*. Sometimes it is a sad little wail.

Breeding August to December. The bulb-shaped nest is made of bark, grasses and root fibres, lined with finer material bound with cobweb and with a side entrance. It is suspended by stranded cobweb from the ceiling of a cave or rock overhang or from man-made structures in dark places. The three eggs are white.

Behaviour The Rock Warbler is a confident ground species. It is able to hop on vertical rock faces and even upside down on overhangs.

Food Ground insects and some seed.

Distribution This bird is restricted to central eastern New South Wales, where it inhabits the sandstones of the geologic Sydney Basin and nearby limestone areas as at Jenolan Caves. It has a preference for rocky outcrops near streams. In more pristine parts of the Blue Mountains it may be the dominant ground species.

Rock Warbler

White-browed Scrub-wren *Sericornis frontalis*

Appearance Length 130mm. Sexes are almost similar. Upper parts are olive-brown shaded to cinnamon. A white eyebrow reaches to the dark brown bill and there is a white spot below the cream eye. The white throat extends in a band towards the neck. The breast is pale and streaked or spotted. The legs are pinkish. The head and face of the female are duller. Immatures are browner than the females.

Voice The call is made up of grating notes and the song is a succession of pleasing *'tsee'* sounds.

Breeding July to December. The nest is made of bark, grasses and leaves. It is dome-shaped with a side entrance and is usually hidden in thick undergrowth on the ground. The two or three whitish eggs are brown freckled.

Behaviour A trusting species, it is generally seen on or near the ground in small parties.

Food Insects and their larvae, spiders and sometimes seed.

Distribution From the coast and adjacent ranges of eastern Australia to Shark Bay in Western Australia. It is equally common in forests and heath provided that there is dense undergrowth. A resident breeder in the Blue Mountains, it is more common than other scrub-wrens which occupy differing habitats.

White-browed Scrub-wren

White-throated Warbler *Gerygone olivacea*

Appearance	Length 110mm. Sexes are similar. The head and upper parts are ash grey with an olive cast. There is a white mark on each side of the forehead. The darker tail is banded black near the tip and shows a white patch when extended in flight. The throat is white and the under parts light yellow. Eyes are red and bill black. Immatures are duller with a yellow throat. They have brown eyes.
Voice	The principal song is unmistakeable. A clear sweet falling and shortly rising chromatic scale is frequently heard. Other calls are softer variations.
Breeding	One or more broods in spring to summer. The nest is a pear-shaped structure of fine bark shreds, bound with cobweb, with a hooded side opening and a dangling tail. There are two or three white to pink eggs splashed darker.
Behaviour	Its presence is revealed by its song as it is an unobtrusive species which forages in the outer foliage, usually in pairs or small groups.
Food	Foliage insects, ants, and other small insects.
Distribution	Northern, eastern and south-eastern Australia. In the Blue Mountains it is a spring migrant from the north, departing again in autumn. Its habitat is open forest but it frequents bushy gardens.

White-throated Warbler

Brown Thornbill *Acanthiza pusilla*

Appearance	Length 100mm. Sexes are similar. Upper parts are an olive-brown with a blackish band near the tip of the tail. There is a light rufous wash on the forehead, upper tail and wings. Under parts are whitish-grey with black markings on the throat and breast. Eyes are red, bill black and legs dark brown. By about a month after birth immatures have acquired their adult plumage.
Voice	The song is varied but consists of a few sweet notes. Other birds are often mimicked. It harshly scolds when alarmed.
Breeding	August to December. The domed nest has an entrance near the top. It is untidily built of grass, fern and bark bound with cobweb close to the ground. The three tiny eggs are white with red-brown freckles.
Behaviour	Though a small bird which spends most of its time actively foraging at lower levels, its inquisitive nature often makes it visible. It may hang upside down to feed. The tail is opened and closed during flight. It often moves in mixed small bird flocks.
Food	Insects from the shrub layer.
Distribution	Widespread in the wetter parts of south-eastern Australia from the coast to the ranges. In the Blue Mountains it is common in areas of dense undergrowth, including gardens.

Brown Thornbill

Striated Thornbill
Acanthiza lineata

Appearance	Length 100mm. Sexes are similar. Upper parts are a pale olive-brown above with a black band near the tip of the tail. There are distinctive white streaks on the head, cheeks and wings. Under parts are a yellowish-white with black streaks. The eyes and bill are dark brown and the legs grey. Immatures are generally darker than the adults.
Voice	The spring song is a soft high-pitched trill. The call is a soft insect-like *'zit'*.
Breeding	July to December. The rounded nest with a small entrance near the top is more neatly constructed than that of the Brown Thornbill. Built of fine grass and bark, bound by cobwebs and lined with soft material, it is suspended in foliage from an outer branch, up to 15m above the ground. The two to four pale pink eggs are spotted red-brown.
Behaviour	This tiny bird gathers in small flocks to feed in the upper foliage of trees. As it hovers to take insects from leaves it fans its tail to show the dark tailband.
Food	Spiders and insects from the upper branches of trees.
Distribution	From the coast to the plateaus and western slopes of south-eastern Australia. In the Blue Mountains it ranges from the rainforest to eucalypt woodland. It is a visitor to parks and gardens.

Striated Thornbill

White-throated Treecreeper *Climacteris leucophaea*

Appearance	Length 160mm. Sexes are similar though the female has a red spot below the cheek. The dark brown of the head shades down the back to brown-grey on the tail. The throat is white and other under parts streaked black and white. The eyes are brown, and the strong legs and feet, and narrow bill are dark brown. Immatures are chestnut rumped.
Voice	The song is a rounded rhythmic one note whistle.
Breeding	August to January. The nest is tucked in a high tree hollow. It is packed up with debris and lined with soft material. The two to four white eggs have brownish-red spots.
Behaviour	Hunts prey on rough-barked trees which are ascended more or less spirally. As it cannot reverse, it then flies to the base of a neighbouring tree for the next ascent. Though generally solitary or in pairs, out of the breeding season it often moves in mixed flocks of small birds.
Food	Mainly ants and insects taken off tree bark.
Distribution	In the forests of the coast and adjacent ranges from south-eastern Queensland to South Australia. It is a sedentary resident breeder in the Blue Mountains. Its range in the upper Blue Mountains overlaps that of the rarer Red-browed Treecreeper, a bird with a rusty-red eyebrow.

White-throated Treecreeper

Red Wattlebird *Anthochaera carunculata*

Appearance	Length 340mm. Sexes are similar. Upper parts are a dark olive-brown with white streaks and the head is dark. Under parts are light brown streaked with white and the belly is yellow. Below the red eye a white triangular patch extends to the black bill. Below this there is a red wattle or cheek flap which lengthens and darkens with age. Legs are pinky-brown.
Voice	Loud harsh coughs.
Breeding	July to December. A rough saucer of sticks and grass, lined with softer material, is built in the fork of a large tree or shrub, 2-12m above the ground. The two to three small pinkish eggs are spotted with red or purple.
Behaviour	A rowdy aggressive bird common in gardens, it is often seen performing aerial acrobatics as it chases insects. It will also hop after them along the ground.
Food	Orb-weaving and other spiders, insects, soft fruit and nectar from a range of native and exotic species.
Distribution	Forests and woodlands from Brisbane to Perth along the southern edge of the continent. Migratory flocks form and leave the breeding grounds in autumn, seeking out flowers in lower altitudes. They are very common in the Blue Mountains where they follow the nectar flow.

Red Wattlebird

Noisy Friarbird *Philemon corniculatus*

Appearance	Length 360mm. Sexes are similar. Head is bare and black and the upper parts shaded brown-grey. Chin is white, the throat and upper breast have long dark-shafted white feathers, and the lower breast and belly are a softer brown-grey. The eyes are red-brown, the black bill is long and curved with a knob and the legs are dark grey. Immatures have greyish heads and the bill knob is small.
Voice	A rackety raucous call is the identifying sound and there are other squawks and harsh notes.
Breeding	August to October. Nest is a capacious cup of bark and grass bound with cobweb and lined with soft material. It is built in the thick foliage of outer branches of trees. The two or three eggs are buff-pink with marbled markings.
Behaviour	A lively and busy bird, it moves in noisy flocks out of the breeding season. It forages through tree foliage where it aggressively defends food resources.
Food	Nectar, native and exotic fruits, and insects.
Distribution	It is widespread in eastern Australia in a variety of habitats from forest to heath and is common in the lower Blue Mountains where it is a resident breeder. It is partially nomadic, following the nectar flow and is regularly seen in suburban gardens.

Noisy Friarbird

Bell Miner

Manorina melanophrys

Appearance	Length 180mm. Sexes are similar. Upper parts are olive-green and under parts pale yellowish-green. Wings and tail are brown. There is an orange triangular spot behind the black eye. Bill and legs are a paler orange.
Voice	The high pitched continuous metallic *'tink-tink'* call is reflected in the alternative common name of Bellbird.
Breeding	July to February. The nest, a roughly constructed cup of twigs and grasses, is woven with cobwebs to supporting branches. The one to three pinkish eggs have darker spots.
Behaviour	The Bell Miner is best known by its distinctive call, for although it lives in large colonies it is well hidden in the dense foliage of trees. An active bird it becomes aggressive in the presence of intruders, flapping its wings, calling harshly and dropping to lower branches or to the ground where the performance is repeated.
Food	Insects and the sugary exudate which covers lerps, a group of eucalypt eating insects.
Distribution	From southern Queensland to Victoria east of the Great Dividing Range. Preferring the damper forests, colonies are found in the Blue Mountains at Bellbird Hill, Kurrajong Heights, along some of the streams draining the Eastern Escarpment, at Springwood and in the Jamison Valley.

Bell Miner

Noisy Miner *Manorina melanocephala*

Appearance Length 280mm. Sexes are similar. The brown-grey upper parts are lightly barred with white and the crown is black. The wings are an olive brown-grey and the brown tail is white-tipped. The grey breast shades whiter towards the belly. There is a bright yellow triangular patch behind the brown eye. The bill and legs are yellow. Upper parts of the immatures are a lighter brown.

Voice Clear piping in various pitches and scolding sounds are generally associated with defence of territories.

Breeding May be at any time but centres on June to December. The cup-shaped nest, made of twigs and grasses with fine lining and sometimes decorated, is suspended from a branch. The two to four cream eggs are spotted red-brown.

Behaviour The birds perform communal activities in a well developed social organisation. All group members assist with the care of the young but the female alone builds the nest and incubates the eggs. Their territory is vigorously defended. The species is active, noisy and aggressive.

Food Fruits, nectar and insects from the ground or foliage.

Distribution Colonies are widespread in the woodlands of eastern Australia. They are sedentary resident breeders in the lower Blue Mountains, favoured by suburbanisation here.

Noisy Miner

Yellow-faced Honeyeater *Lichenostomus chrysops*

Appearance Length 180mm. Sexes are similar, olive-brown above and grey below. The yellow band which extends from behind and below the blue-grey eye to the steel grey bill is bordered above and below by a black band. Legs are grey. Immatures are duller.

Voice Sweet chattering.

Breeding Late spring or summer. The nest, a neat cup made of grass and bark bound with cobweb, lined with finer material and sometimes camouflaged with lichen, is suspended from a horizontal fork. The two or three pink to buff eggs are spotted darker.

Behaviour This is a lively bird which moves singly, in pairs or in flocks, sometimes of mixed species. It is known to alight on humans and animals seeking hair or fur for nest material.

Food Nectar from native and introduced flowers, honeydew, fruit and foliage insects.

Distribution Forests and woodlands of the ranges of eastern Australia. South of Brisbane its range extends to the coast. It is one of the more common honeyeaters of the Blue Mountains. There is a marked northern migration in autumn but some birds overwinter and may visit gardens for nectar and fruit.

Yellow-faced Honeyeater

White-eared Honeyeater *Lichenostomus leucotis*

Appearance	Length 200mm. Sexes are similar. It is olive-green above with a grey streaked crown. The face and throat are black. Under parts are a lighter green-yellow. A white ear patch extends below the light brown eye. The bill and legs are black. Immatures are duller with an olive-green crown.
Voice	Resonant short calls which are seasonally variable.
Breeding	August to December. It broods later and more than once if the season is suitable. The cup-shaped nest of bark and grasses is lined with softer material including fur and hair. The two or three eggs are white with faint reddish spots.
Behaviour	This honeyeater is very tame in the breeding season. After keeping lookout on a high bare branch, individuals will alight on humans and animals to secure hair, fibre or fur for nesting material. As it has only a short bill, it obtains nectar by piercing the base of bell-shaped flowers from the outside, thus bypassing its role as a transferer of pollen.
Food	The sap of angophoras and eucalypts, and nectar from shallow flowers are its main foods. Insects are also foraged from eucalypts.
Distribution	With a range through south-eastern and south-western Australia, it is a sedentary resident breeder in the heaths, woodlands and forests of the Blue Mountains.

White-eared Honeyeater

White-naped Honeyeater
Melithreptus lunatus

Appearance	Length 140mm. Sexes are similar. Upper parts are olive and the under parts white. The head is black with a fine white crescent on the nape. The bare skin around the brown eye is red. The bill is black and the legs brown. Immatures are duller with a brown crown, buff crescent-shaped band and whitish eye patch.
Voice	Grating whistle, bright chattering and *'tsit'* sounds.
Breeding	July to November. The nest is a compact cup of bark and grass, lined with fine material and suspended from fine drooping branchlets in the outer foliage. The two or three pale buff eggs are spotted red-brown.
Behaviour	The White-naped Honeyeater lives in flocks and forages acrobatically in the higher foliage.
Food	Foliage insects and larvae, nectar and insects foraged from blossoms, manna and honeydew.
Distribution	Eastern and south-western Australia in eucalypt forests but it also follows nectar into the rich heathlands. In the Blue Mountains there is some local altitudinal movement with overwintering in the lower Mountains but generally there is a marked northward migration. It is present in numbers in the upper Mountains when heaths such as The Pinnacles near Mt Hay, and Narrow Neck are in flower.

White-naped Honeyeater

Crescent Honeyeater   *Phylidonyris pyrrhoptera*

Appearance	Length 155mm. Upper parts of the male are black with elongated yellow patches on the wings and tail. The white breast is divided down each side by a black crescent. The belly is grey. Upper parts and wing patches of the female are olive-brown and the breast crescent is indistinct. The eyes are dark red, the bill black and the legs grey. Immatures lack the crescent and are duller than the adults.
Voice	The call is a high pitched *'e-gypt, e-gypt'*.
Breeding	July to January. The nest is a deep bulky cup of twigs and bark lined with softer material, built in the fork of a shrub up to 2m above the ground. The two or three pinkish eggs are spotted with browns and reds towards the larger end.
Behaviour	When food supplies are plentiful the Crescent Honeyeater forms small groups to forage in the middle layers of the forest where it is more often heard than seen.
Food	Nectar of banksias and members of the Epacridaceae family, insects and sap.
Distribution	From just north of Sydney to the Mt Lofty Ranges in South Australia, and in Tasmania. In winter it inhabits coastal areas and moves to higher altitudes to breed. In the Blue Mountains it prefers the moister valleys but follows the nectar flow.

Male Crescent Honeyeater

New Holland Honeyeater *Phylidonyris novaehollandiae*

Appearance	Length 175mm. Though the sexes are similar the male is slightly larger. It is overall black and white streaked with yellow on the edges of the wings and tail. The eyes, ear tufts, and sparse 'beard' are white. The long slender bill and legs are black. Immatures have dark eyes and are streaked brown and grey.
Voice	Quick high-pitched loud whistles, harsh chattering and scolding.
Breeding	Mainly March to May and August to October. The cup-shaped nest of tightly woven bark and grass is lined with softer material. Its position depends on the season. In summer it is shaded by foliage while in winter it is placed in a sunny position on the outer branches of a large shrub. The one to three dull pink eggs are red spotted.
Behaviour	This bird frequently gathers in colonies in flowering trees. Being very active, it engages in rapid chasing or joins with a small party to harass sleeping owls.
Food	Insects taken on the wing, sometimes several at a time, sap, and nectar from grevilleas, banksias and eucalypts.
Distribution	South-eastern and south-western fringes of Australia from heaths to the edges of forest, following the nectar flow. It is a resident breeder in the Blue Mountains.

New Holland Honeyeater

Eastern Spinebill
Acanthorhynchus tenuirostris

Appearance	Length 160mm. Sexes are similar though the female is duller. Upper parts are a grey-brown. The black of the head runs to separate the white throat from the rufous belly. There is a tawny patch on the throat. The outer tail feathers are white. The long curved bill is black, the eyes red and the legs dark brown. Immatures are olive-grey above and fawn below.
Voice	A clear shrill *'tink'* rapidly repeated about six or seven times and several other musical calls.
Breeding	Through the warmer months. More than one brood is produced in a suitable season. Nest is a cup made from plant fibres and grass, lined with softer material and attached to branchlets. The two or three eggs are fawn with darker spots.
Behaviour	The Eastern Spinebill is an active bird whose wings make a flipping sound during its distinctive fast darting flight. It hovers to take nectar from hanging flowers.
Food	Nectar and insects.
Distribution	Eastern and south-eastern Australia in forest, woodland and heath, particularly where eucalypts and banksias abound. It is a commonly seen and heard breeding resident in the Blue Mountains where it follows the nectar flow.

Eastern Spinebill

Spotted Pardalote
Pardalotus punctatus

Appearance	Length 90mm. The crown, forehead, wings and tail of the male are black with tiny white spots. The back is brown with buff spots. A rich chestnut-brown line extends from the rump towards the tail. There is a white eyebrow above the grey eye. Sides of the face are grey, the throat and undertail feathers are bright yellow and the belly buff. The bill is black and the legs light brown. Females are duller, lack the yellow throat and have cream spotted foreheads. Immatures are similar to females but have greyish crowns.
Voice	A high beautifully clear three or four note call is heard constantly during the breeding season.
Breeding	October to January. A dome-shaped nest of fine material is built in a nesting burrow excavated in a creek bank, cliff or garden sand heap. The three to five tiny eggs are white.
Behaviour	Though generally single or in pairs it will join with many others in a mixed flock to forage through foliage. Breeding pairs perform wing-spreading displays at the entrance to the nesting burrow.
Food	Spiders, caterpillars, beetles, moths and scale insects.
Distribution	Eastern and south-western Australia. In the Blue Mountains they live in a range of habitats from the tallest forests to heaths, parks and gardens.

Male Spotted Pardalote

Silvereye *Zosterops lateralis*

Appearance	Length 120mm. Sexes are similar. Though there is a wide variation, the overall colour is yellow-green with a grey-shaded breast. The throat is sometimes bright yellow. There is a conspicuous white feathered ring around the eye.
Voice	The song is a musical chirping and warbling. The call is an extended *'cheeu'*. It is capable of mimicry.
Breeding	August to January. There is more than one brood in a suitable season. The cup-shaped nest of grasses, hair, fur, cobweb and finer lining, is suspended among branchlets in outer foliage, a few metres above the ground. The two or three eggs are a pale green-blue.
Behaviour	Out of the breeding season it moves in mixed small bird flocks foraging through shrubs and the middle to lower foliage of trees. It moves rapidly through the trees calling constantly.
Food	Insects, worms, fruit and nectar.
Distribution	From Cape York, Queensland, around the southern coastal fringe to Shark Bay, Western Australia. Those of Tasmania are migratory. It is a sedentary breeder in the Blue Mountains where it frequents every habitat except rainforest and commonly forages through gardens, particularly where there are fruit trees.

Silvereye

Red-browed Firetail *Emblema temporalis*

Appearance	Length 120mm. Sexes are similar. Head and under parts are grey, back and wings olive-green and the rump and upper tail red. There is a broad red band above the maroon eye. The red-sided bill is black above and below. The legs are buff. Immatures lack the red on the eyebrow and bill.
Voice	A high pitched *'seee'*.
Breeding	September to December or later. Nest is bulky and flask shaped, made of dried and green grasses with some coarser material and lined with finer material. It has a tunnel entrance and is set in thick shrubs. The four to eight eggs are small and white.
Behaviour	This Firetail has an interesting courtship ritual. The male, holding a grass stem in his bill, jumps stiffly towards the female. Then, when she tosses back her head, he wipes his beak and sings. During the non-breeding season it forms small close flocks which roost together. It seeks food close to cover from which sorties are made.
Food	Grass and introduced weed seed, berries and insects.
Distribution	Along the coastal belt from North Queensland to Adelaide. It has been introduced to Perth. A breeding resident in the Blue Mountains, it lives in most habitats where there are clearings adjacent to cover and has adapted well to gardens.

Red-browed Firetail

Olive-backed Oriole
Oriolus sagittatus

Appearance	Length 260mm. Sexes are similar but the female is duller. Upper parts are olive with the wing quills and tail dark with pale edges. Under parts are white with strong black streaks. The eye is orange-red, the long bill orange-brown and the legs steel grey. Immatures are duller with rufous edging to the wings and tail.
Voice	A distinctive rolling warble often heard in the early morning hours. It may mimic other birds.
Breeding	September to January. Nest is a tenuous but cohesive cup of bark, wool, leaves and grass attached to outer branchlets. The two to four cream eggs are grey and brown spotted.
Behaviour	The presence of this unobtrusive arboreal bird is often revealed by its characteristic song. Its undulating flight is usually short.
Food	Mainly small native fruits, insects, and small exotic fruits where available.
Distribution	From Adelaide along the eastern and northern parts of the continent to the Kimberleys in Western Australia where woodland, eucalypt forest and edges of rainforest are favoured. It is a summer migrant in the Blue Mountains but some birds overwinter here. It is more common in the middle and lower Mountains.

Olive-backed Oriole

# Satin Bower Bird	*Ptilonorhynchus violaceus*

Appearance	Length 270-330mm. The male is blue-black overall with a sapphire eye, stubby dull blue bill and pale legs and feet. In the female the head and upper parts are grey-green, the wings olive-brown, washed gold beneath and the under parts creamy with olive-brown scallops. The eyes are sapphire, bill dull grey and the feet and claws pale grey. Immatures generally resemble the adult female. Adult plumage is gained at six to seven years.

Voice	Great variety including whirring, rattling and creaking sounds. Contact call is a confident *'cheeru'*. During display there are rapid repeated churrings and local sound mimicry.

Breeding	September to January. The shallow nest of twigs lined with dried leaves is often high in a tree. The two or three dark cream eggs are darker marked.

Behaviour	Dominant males individually make and maintain U-shaped bowers of pliant twigs and sedges, decorated with blue and yellow pieces, as display courts to which many females are enticed to mate. There is some seasonal clan movement.

Food	Grass, leaves, fruit, insects and garden vegetables.

Distribution	Coastal eastern Australia. It is common in the Blue Mountains where it is a breeding resident. Though forests are prefered bowers are often set in bushy gardens.

Male Satin Bower Bird

Grey Butcherbird
Cracticus torquatus

Appearance Length 300mm. Sexes are similar. The head is black with a white shadow in front of the eyes. Upper parts are dark grey with a white collar. The tail is black with a white tip. Throat is white and the under parts pale grey. The grey bill has a black hooked tip. Eyes are dark brown and the legs steel grey. Immatures to two years are brown and white.

Voice A clear rollicking song and a variety of other musical notes. Male and female sometimes duet. It has a mimicing ability. The song is heard more in spring and autumn.

Breeding July to January. The nest is a neat cup made of twigs lined with finer material and placed in a high vertical fork. The three or four eggs are variably grey, blue or olive to brown with red to brown markings.

Behaviour It carries food with its hooked bill as the feet and legs are too small to grapple prey. Food is then secured for better manipulation, often in the fork of a branch, and torn apart with the bill.

Food Preys on ground insects and small animals.

Distribution Almost Australia wide, it is a resident in the forested areas of the Blue Mountains and is particularly common in the middle and lower Mountains. It is easily attracted to gardens by offerings of suitable food.

Grey Butcherbird

Australian Magpie *Gymnorhina tibicen*

Appearance Length 440mm. The male is glossy black with white on the nape, rump and under tail. The white tail is edged in black and the black wings have a broad white band. Nape and rump of the female are greyer. Eyes are red-brown, the bill bluish with a black tip and the legs black. For about the first year immatures are mottled grey.

Voice The territorial call, best heard during spring, is a rich carolling, often by several birds at once. The call is harsh and aggressive during invasion of its territory. In summer the persistent begging of young is a common call.

Breeding August to October. The bowl-shaped nest of sticks with finer lining is placed in a fork or the outer branches of a tree some distance from the ground. The three to five eggs are grey to blue-green with brown spots.

Behaviour Magpies live in a family group of up to ten birds, each of which aggressively defends the territory. People are a frequent target during the breeding season.

Food Worms, insects and fruit usually foraged from the ground.

Distribution Common throughout Australia except for the northern extremities. Resident breeders in the Blue Mountains, they inhabit timbered areas but have adapted to settlement and are common visitors to gardens.

Australian Magpie

Pied Currawong

Strepera graculina

Appearance	Length 470mm. Sexes are similar. It is black with large white patches on the tip of the tail, towards the rump and the top of the wings. The eyes are yellow, and the large sharp bill and the legs are black. Immatures tend to be an ash grey-black with fewer white patches.
Voice	Utters a number of calls, often in flight, including a long loud *'curra-wong, curra-wong'* and a rising then falling *'we-ee-oo'*. The carolling of flocks resounds through the Blue Mountains from autumn to winter.
Breeding	Spring. The flattish nest of sticks lined with grass and bark is placed high in a eucalypt. The three small oval eggs are light brown with darker patches. Channel-billed Cuckoos sometimes lay an egg in these nests with their large young being raised by the Currawongs.
Behaviour	Alert, ever watchful birds, they feed on the ground and in larger trees, swooping at intruders. In winter they form great flocks, roosting communally in valleys at night. Single pairs then disperse for the breeding season.
Food	Grubs, insects including female stick insects, and fruit.
Distribution	Widespread in the eucalypt forests from Queensland to Victoria. It is one of the most frequently seen and heard birds in the towns and forests of the Blue Mountains.

Pied Currawong

Grey Currawong *Strepera versicolor*

Appearance Length 500mm. Sexes are similar. Males are an overall grey, with the under tail and tail base white. Eyes are yellow and the powerful bill and the legs are black. The females are smaller, duller and have brown eyes. Immatures are duller than the adults.

Voice The flight call is a *'ching-ching'*. While perched it may utter many small unusual notes.

Breeding July to November. The large shallow bowl-shaped nest of sticks lined with coarse plant fibre is placed well above the ground in the fork of a tree. The two or three eggs are buff with heavy markings of brown and grey.

Behaviour Though generally solitary or in pairs, out of the breeding season this bird forms small family groups. Food is sought from the ground or trees and it will prise off bark with its powerful bill to obtain insects. It is harassed by resident Pied Currawongs if it ventures into the latters' territory.

Food Insects, larvae, nestlings, and fruit raided from orchards.

Distribution Southern parts of the continent from the coast to the semi-arid areas. It is a sedentary breeder in the forests and woodlands of the Blue Mountains though it sometimes migrates to lower more open country in winter. It will move into gardens for ripe fruit.

Grey Currawong

Australian Raven *Corvus coronoides*

Appearance Length 520mm. Sexes are similar. The glossy black plumage has a purple-green sheen. Throat feathers are long and full. Eyes are white, the large powerful bill is black and the legs are dark brown. Immatures have duller plumage and brown eyes.

Voice The loud *'aah-aaah'*, dying in a descending wail is most frequently made in flight and during the morning as they daily confirm their territories.

Breeding Spring. The large deep nest, made of sticks lined with grass and wool, is placed in a high tree. The four to six eggs are pale green, blotched with olive-brown. Australian Ravens mate for life.

Behaviour This is a wary bird with a heavy measured flight and a flat-footed walk. In the Blue Mountains it moves in small flocks and is seldom solitary.

Food Insects, carrion and spilt grain.

Distribution Eastern half and the south-west of Australia. Open country or open forest with good overviews are necessary to this species. It is a breeding resident in the Blue Mountains. Widespread but not particularly common here, it has benefited from settlement which has created open space and garbage tips where it forages.

Australian Raven

Checklist of Birds of the Blue Mountains

Grebes
Australasian Grebe — *Tachybaptus novaehollandiae*
Pelicans and Allies
Australian Pelican — *Pelecanus conspicillatus*
Great Cormorant — *Phalacrocorax carbo*
Pied Cormorant — *P.varius*
Little Black Cormorant — *P.sulcirostris*
Little Pied Cormorant — *P.melanoleucos*
Herons and Allies
Pacific Heron — *Ardea pacifica*
White-faced Heron — *A.novaehollandiae*
Great Egret — *Egretta alba*
Little Egret — *E.garzetta*
Rufous Night Heron — *Nycticorax caledonicus*
Duck-like Birds
Sacred Ibis — *Threskiornis aethiopica*
Straw-necked Ibis — *T.spinicollis*
Black Swan — *Cygnus atratus*
Pacific Black Duck — *Anas superciliosa*
Mallard* — *A.platyrhynchos**
Grey Teal — *A.gibberifrons*
Maned Duck — *Chenonetta jubata*
Birds of Prey
Black-shouldered Kite — *Elanus notatus*
Pacific Baza — *Aviceda subcristata*
Black Kite — *Milvus migrans*
Whistling Kite — *Haliastur sphenurus*
Brown Goshawk — *Accipiter fasciatus*
Collared Sparrowhawk — *A.cirrhocephalus*
Grey Goshawk — *A.novaehollandiae*
White-bellied Sea Eagle — *Haliaeetus leucogaster*
Wedge-tailed Eagle — *Aquila audax*
Little Eagle — *Hieraaetus morphnoides*
Marsh Harrier — *Circus aeruginosus*
Peregrine Falcon — *Falco peregrinus*
Australian Hobby — *F.longipennis*
Brown Falcon — *F.berigora*
Australian Kestrel — *F.cenchroides*
Button Quails and Allies
Brown Quail — *Coturnix australis*
Painted Button Quail — *Turnix varia*
Little Button Quail — *T.velox*
Red-chested Button Quail — *T.pyrrhothorax*
Buff-banded Rail — *Rallus philippensis*
Lewin's Rail — *R.pectoralis*
Dusky Moorhen — *Gallinula tenebrosa*
Purple Swamphen — *Porphyrio porphyrio*
Eurasian Coot — *Fulica atra*
Waders and Gulls
Masked Lapwing — *Vanellus miles*
Black-fronted Plover — *Charadius melanops*

Latham's Snipe	*Gallinago hardwickii*
Silver Gull	*Larus novaehollandiae*
Whiskered Tern	*Chlidonias hybrida*

Pigeons and Doves

Domestic Pigeon*	*Columba livia**
Spotted Turtledove*	*Streptopelia chinensis**
Brown Cuckoo-dove	*Macropygia amboinensis*
Peaceful Dove	*Geopelia placida*
Diamond Dove	*G. cuneata*
Common Bronzewing	*Phaps chalcoptera*
Brush Bronzewing	*P. elegans*
Crested Pigeon	*Ocyphaps lophotes*
Wonga Pigeon	*Leucosarcia melanoleuca*

Cockatoos and Parrots

Glossy Black Cockatoo	*Calyptorhynchus lathami*
Yellow-tailed Black Cockatoo	*C. funereus*
Gang-gang Cockatoo	*Callocephalon fimbriatum*
Galah	*Cacatua roseicapilla*
Little Corella	*C. sanguinea*
Sulphur-crested Cockatoo	*C. galerita*
Rainbow Lorikeet	*Trichoglossus haematodus*
Musk Lorikeet	*Glossopsitta concinna*
Little Lorikeet	*G. pusilla*
Australian King Parrot	*Alisterus scapularis*
Crimson Rosella	*Platycercus elegans*
Eastern Rosella	*P. eximius*
Red-rumped Parrot	*Psephotus haematonotus*

Cuckoos

Pallid Cuckoo	*Cuculus pallidus*
Brush Cuckoo	*C. variolosus*
Fan-tailed Cuckoo	*C. pyrrhophanus*
Shining-bronze Cuckoo	*Chrysococcyx lucidus*
Common Koel	*Eudynamis scolopacea*
Channel-billed Cuckoo	*Scythrops novaehollandiae*
Pheasant Coucal	*Centropus phasianinus*

Owls

Powerful Owl	*Ninox strenua*
Boobook Owl	*N. novaeseelandiae*
Barking Owl	*N. connivens*
Barn Owl	*Tyto alba*

Frogmouths and Nightjars

Tawny Frogmouth	*Podargus strigoides*
Owlet Nightjar	*Aegotheles cristatus*
White-throated Nightjar	*Caprimulgus mystacalis*

Swifts

White-throated Needletail	*Hirundapus caudacutus*
Fork-tailed Swift	*Apus pacificus*

Kingfishers and Allies

Azure Kingfisher	*Ceyx azurea*
Laughing Kookaburra	*Dacelo novaeguineae*
Sacred Kingfisher	*Halcyon sancta*
Rainbow Bee-eater	*Merops ornatus*
Dollar Bird	*Eurystomus orientalis*

Perching Birds

Superb Lyrebird	*Menura novaehollandiae*
Welcome Swallow	*Hirundo neoxena*
Tree Martin	*Cecropis nigricans*
Fairy Martin	*C.ariel*
Richard's Pipit	*Anthus novaeseelandiae*
Black-faced Cuckoo-shrike	*Coracina novaehollandiae*
White-bellied Cuckoo-shrike	*C.papuensis*
Cicada Bird	*C.tenuirostris*
White-winged Triller	*Lalage sueurii*
Red-whiskered Bulbul*	*Pyconotus jocosus* *
White's Thrush	*Zoothera dauma*
Blackbird	*Turdus merula*
Rose Robin	*Petroica rosea*
Flame Robin	*P.phoenicea*
Scarlet Robin	*P.multicolor*
Red-capped Robin	*P.goodenovii*
Eastern Yellow Robin	*Eopsaltria australis*
Jacky Winter	*Microeca leucophaea*
Crested Shrike-tit	*Falcunculus frontatus*
Golden Whistler	*Pachycephala pectoralis*
Rufous Whistler	*P.rufiventris*
Grey Shrike-thrush	*Colluricincla harmonica*
Black-faced Monarch	*Monarcha melanopsis*
Leaden Flycatcher	*Myiagra rubecula*
Satin Flycatcher	*M.cyanoleuca*
Restless Flycatcher	*M.inquieta*
Rufous Fantail	*Rhipidura rufifrons*
Grey Fantail	*R.fuliginosa*
Willie Wagtail	*R.leucophrys*
Eastern Whipbird	*Psophodes olivaceus*
Spotted Quail-thrush	*Cinclosoma punctatum*
Rufous Songlark	*Cinclorhamphus mathewsi*
Superb Fairy-wren	*Malurus cyaneus*
Variegated Fairy-wren	*M.lamberti*
Southern Emu-wren	*Stipiturus malachurus*
Pilot Bird	*Pycnoptilus floccosus*
Rock Warbler (Origma)	*Origma solitaria*
Large-billed Scrub-wren	*Sericornis magnirostris*
Yellow-throated Scrub-wren	*S.citreogularis*
White-browed Scrub-wren	*S.frontalis*
Chestnut-rumped Hylacola	*S.pyrrhopygius*
Brown Gerygone	*Gerygone mouki*
White-throated Warbler	*G.olivacea*
Brown Thornbill	*Acanthiza pusilla*
Buff-rumped Thornbill	*A.reguloides*
Yellow-rumped Thornbill	*A.chrysorrhoa*
Yellow Thornbill	*A.nana*
Striated Thornbill	*A.lineata*
Varied Sittella	*Daphoenositta chrysoptera*
White-throated Treecreeper	*Climacteris leucophaea*
Red-browed Treecreeper	*C.erythrops*
Brown Treecreeper	*C.picumnus*

Common Name	Scientific Name
Red Wattlebird	*Anthochaera carunculata*
Little Wattlebird	*A.chrysoptera*
Noisy Friarbird	*Philemon corniculatus*
Little Friarbird	*P.citreogularis*
Regent Honeyeater	*Xanthomyza phrygia*
Bell Miner	*Manorina melanophrys*
Noisy Miner	*M.melanocephala*
Lewin's Honeyeater	*Meliphaga lewinii*
Yellow-faced Honeyeater	*Lichenostomus chrysops*
White-eared Honeyeater	*L.leucotis*
Yellow-tufted Honeyeater	*L.melanops*
Fuscous Honeyeater	*L.fuscus*
White-plumed Honeyeater	*L.penicillatus*
Brown-headed Honeyeater	*Melithreptus brevirostris*
White-naped Honeyeater	*M.lunatus*
Crescent Honeyeater	*Phylidonyris pyrrhoptera*
New Holland Honeyeater	*P.novaehollandiae*
White-cheeked Honeyeater	*P.nigra*
Tawny Crowned Honeyeater	*P.melanops*
Eastern Spinebill	*Acanthorhynchus tenuirostris*
Scarlet Honeyeater	*Myzomela sanguinolenta*
Mistletoe Bird	*Dicaeum hirundinaceum*
Spotted Pardalote	*Pardalotus punctatus*
Striated Pardalote	*P.striatus*
Silvereye	*Zosterops lateralis*
European Goldfinch*	*Carduelis carduelis**
House Sparrow	*Passer domesticus**
Red-browed Firetail	*Emblema temporalis*
Beautiful Firetail	*E.bella*
Zebra Finch	*Poephila guttata*
Double-barred Finch	*P.bichenovii*
Common Starling*	*Sturnus vulgaris**
Common Mynah*	*Acridotheres tristis**
Olive-backed Oriole	*Oriolus sagittatus*
Satin Bower Bird	*Ptilonorhynchus violaceus*
Green Catbird	*Ailuroedus crassirostris*
White-winged Chough	*Corcorax melanorhamphos*
Australian Magpie-lark	*Grallina cyanoleuca*
White-breasted Woodswallow	*Artamus leucorhynchus*
Masked Woodswallow	*A.personatus*
Dusky Woodswallow	*A.cyanopterus*
Grey Butcherbird	*Cracticus torquatus*
Australian Magpie	*Gymnorhina tibicen*
Pied Currawong	*Strepera graculina*
Grey Currawong	*S.versicolor*
Australian Raven	*Corvus coronoides*
Little Raven	*C.mellori*

Introduced into Australia

INDEX

Acanthiza lineata	38	Lyrebird, Superb		20
pusilla	37	Magpie, Australian		56
Acanthorhynchus tenuirostris	49	*Malurus cyaneus*		32
Alisterus scapularis	12	*lamberti*		33
Anthochaera carunculata	40	*Manorina melanocephala*		43
Aquila audax	5	*melanophrys*		42
Bower Bird, Satin	54	*Melithreptus lunatus*		46
Bronzewing, Brush	7	*Menura novaehollandiae*		20
Butcherbird, Grey	55	Miner, Bell		42
Cacatua galerita	11	Noisy		43
roseicapilla	10	*Ninox novaeseelandiae*		16
Callocephalon fimbriatum	9	*Origma solitaria*		34
Calyptorhynchus funereus	8	Oriole, Olive-backed		53
Chenonetta jubata	4	*Oriolus sagittatus*		53
Climacteris leucophaea	39	Owl, Boobook		16
Cockatoo, Gang-gang	9	*Pachycephala pectoralis*		26
Sulphur-crested	11	*rufiventris*		27
Yellow-tailed Black	8	Pardalote, Spotted		50
Colluricincla harmonica	28	*Pardalotus punctatus*		50
Coracina novaehollandiae	22	*Petroica multicolor*		24
Corvus coronoides	59	*rosea*		23
Cracticus torquatus	55	*Phaps elegans*		7
Cuckoo, Fan-tailed	14	*Philemon corniculatus*		41
Channel-billed	15	*Phylidonyris novaehollandiae*		48
Cuckoo-shrike, Black-faced	22	*pyrrhoptera*		47
Cuculus pyrrhophanus	14	*Platycercus elegans*		13
Currawong, Grey	58	*Podargus strigoides*		17
Pied	57	*Psophodes olivaceus*		31
Dacelo novaeguineae	18	*Ptilonorhynchus violaceus*		54
Duck, Maned	4	Raven, Australian		59
Eagle, Wedge-tailed	5	*Rhipidura fuliginosa*		30
Emblema temporalis	52	*rufifrons*		29
Eopsaltria australis	25	Robin, Rose		23
Fairy-wren, Superb	32	Scarlet		24
Variegated	33	Eastern Yellow		25
Falco berigora	6	Rosella, Crimson		13
Falcon, Brown	6	Scrub-wren, White-browed		35
Fantail, Grey	30	*Scythrops novaehollandiae*		15
Rufous	29	*Sericornis frontalis*		35
Firetail, Red-browed	52	Shrike-thrush, Grey		28
Friarbird, Noisy	41	Silvereye		51
Frogmouth, Tawny	17	Spinebill, Eastern		49
Galah	10	*Strepera graculina*		57
Gerygone olivacea	36	*versicolor*		58
Gymnorhina tibicen	56	Swallow, Welcome		21
Halcyon sancta	19	Thornbill, Brown		37
Hirundo neoxena	21	Striated		38
Honeyeater, Crescent	47	Treecreeper, White-throated		39
New Holland	48	Warbler, Rock		34
White-eared	45	White-throated		36
White-naped	46	Wattlebird, Red		40
Yellow-faced	44	Whipbird, Eastern		31
King Parrot, Australian	12	Whistler, Golden		26
Kingfisher, Sacred	19	Rufous		27
Kookaburra, Laughing	18	*Zosterops lateralis*		51
Lichenostomus chrysops	44			
leucotis	45			